Chemistry for the gifted and talented

Written by Tim Jolliff

RSC Schoolteacher Fellow 2005–2006

Chemistry for the gifted and talented

Written by Tim Jolliff

Edited by Colin Osborne, John Johnston and Gemma Tobiasen

Designed by Russel Spinks

Published and distributed by the Royal Society of Chemistry

Printed by the Royal Society of Chemistry

Copyright © Royal Society of Chemistry 2007

Registered charity No. 207980

For further information on other educational activities undertaken by the Royal Society of Chemistry write to:

Education Department
Royal Society of Chemistry
Burlington House
Piccadilly
London W1J 0BA

Information on other Royal Society of Chemistry activities can be found on its websites:
www.rsc.org
www.chemsoc.org/LearnNet contains resources for teachers and students from around the world.
www.chemistryteachers.org contains curriculum linked resources.
ISBN-10: 0-85404-288-1
ISBN-13: 978-0-85404-288-3

British Library Cataloguing in Publication Data.

A catalogue for this book is available from the British Library.

Foreword

The Royal Society of Chemistry is passionately interested in getting gifted and talented students interested in the further study of the chemical sciences. The RSC has thus produced this volume of activities to interest, challenge and excite young people in the subject. The activities are designed to be used either with or without teacher supervision and to extend even the most able student.

Professor Jim Feast CChem FRSC
President, Royal Society of Chemistry

Acknowledgements

The production of these resources was only made possible by the advice and assistance of a large number of people. The author and the RSC express their gratitude to the following people and everyone who was involved with the project.

Chris Foster, King's College, Taunton, who selflessly lent out his office for the year and provided much inspiration for the activities.
Colin Osborne, Education Manager, Schools and Colleges, Royal Society of Chemistry and Maria Pack, Assistant Education Manager, Schools and Colleges, Royal Society of Chemistry who both patiently proof read and corrected the spelling, punctuation, grammar and chemistry.
Kay Stephenson, Assistant Education Manager, Schools and Colleges, Royal Society of Chemistry
Tim Greene, Clifton College, Bristol
Tim Harrison, Bristol University
Ian Stanbridge, Colyton Grammar School, Devon
Chas McCaw, Winchester College, Winchester
Peter Hollamby, Cardiff
Peter Borrows, CLEAPSS, Brunel University
Steve Lewis, Shrewsbury Sixth Form College, Shrewsbury
Andrew Haigh, Loughborough Grammar School, Loughborough
John Holman, National Science Learning Centre, York
Anthony Hardwick, Dr Challoner's Grammar School, Amersham
Lesley Stanbury, St Albans School, St Albans
Judith Edge, Holmes Chapel Comprehensive School, Cheshire
Neil Dixon, South Bromsgrove High School, Worcestershire
Jeff Hancock, formerly of King Edward's School, Birmingham
Peter Wothers, University of Cambridge
James Keeler, University of Cambridge
Keith Taber, University of Cambridge
Peter Dossett, King's College, Taunton
Jonathan Clayden, Manchester University
Emma Chedzoy, King's College, Taunton
Library and Information Centre, Royal Society of Chemistry
Barry Meatyard, National Academy of Gifted and Talented Youth, Warwick University
Pat O'Brien, Educational Consultant, Kent
Bob Mudd, Bridgewater College, Somerset
Ted Lister, Leamington Spa
Sarah Codrington, Nuffield Curriculum Centre, London
Vicki Barwick, LGC, Teddington

The author would particularly like to thank the Headmasters of Queen's College, Taunton for agreeing to the year's secondment, and King's College, Taunton for kindly providing office facilities.

Contents

Preface

I passionately believe that adapting our teaching to challenge and stimulate the most able students can have huge benefits. My own chemistry teacher's enthusiasm to stretch his  gifted students inspired me to both study and teach chemistry. I treasure the hope that this publication will help teachers differentiate in their lessons for the most able students of all ages and that students will find their interest grows into fascination for a subject that offers, unlike some other subjects, real intellectual rewards.

Tim Jolliff
Royal Society of Chemistry Schoolteacher Fellow 2005–2006

Introduction

Clarification of terms

Giftedness in individuals can be used to describe adults or children that have a capacity for high levels of expertise. Giftedness in children could be thought of as 'expertise in development'.[1] They may or may not be currently high achievers and their potential may emerge at any stage in their development. The Department for Children, Schools and Families (DCSF) uses the term *gifted* to mean the most able in academic subjects like English, maths, history *etc*. The term *talented* has come to refer to those with abilities in areas such as sport, art, music or drama. The gifted and talented students can be thought of as the top fraction of the ability range. Young Gifted & Talented (YG&T) – the new national programme for gifted and talented education – encourages the top 10 per cent to join their membership. Other organisations also take the top 10 per cent as gifted and talented. *Severely gifted* children are exceptionally able and represent a much smaller fraction of the population.

Why something special should be done for the most able students

It is axiomatic in education that *every child matters*. All students deserve an education that gives them the opportunity to maximise their potential. Children need to be challenged by tasks that take into account their abilities and prior knowledge. Gifted provision in the classroom is about meeting the individual needs of the most able along with the other students. Able students, in common with others, can get bored and disaffected by lessons that have an inadequate level of demand or interest for them. Some gifted students become difficult and uncooperative in lessons where they have felt de-motivated by the lack of demand in the tasks set.

Some people involved in education are wary of special provision for the most able. Among their concerns is that such provision may be elitist and lead to lower ability students being less well provided for. They are likely to comment that, if an activity for gifted students is worthwhile, it should be made available to all students. Gifted provision is not elitist – it is not motivated by a belief that some students matter more than others, but by a belief that what is needed to meet the needs of some students differs to that which is needed for others.

Contrary to the belief that the other students suffer when provision is made for gifted students, there is evidence from OFSTED that when the needs of the able students are met, standards of achievement are raised for all students.[2]

Overview of talented and gifted provision

The mind maps below summarise some of the generic topics discussed in the literature on gifted and talented education. There are three maps shown for clarity, but they really form one large whole.

Administration

Schools in England are expected to have produced policies for the gifted and talented and to have put them into practice. Further information about such generic issues is available from *www.standards.dfes.gov.uk/giftedandtalented/* (accessed April 2007).

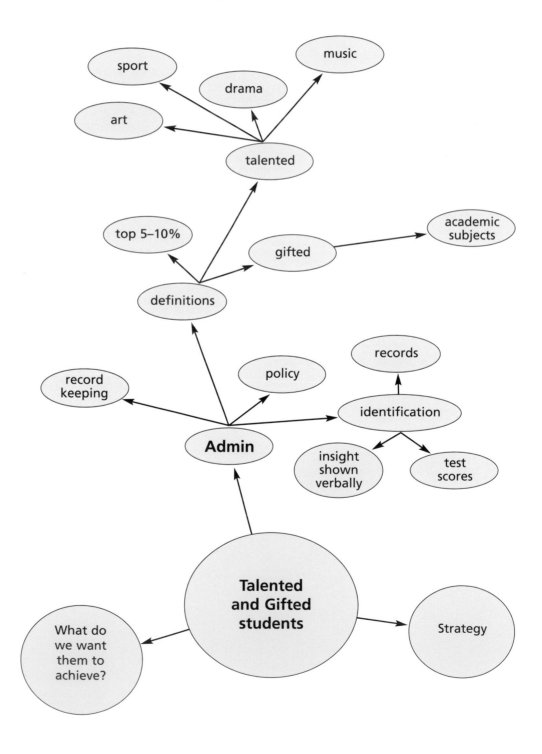

Strategy

This includes an overview of the types of additional and alternative learning experiences that students can benefit from.

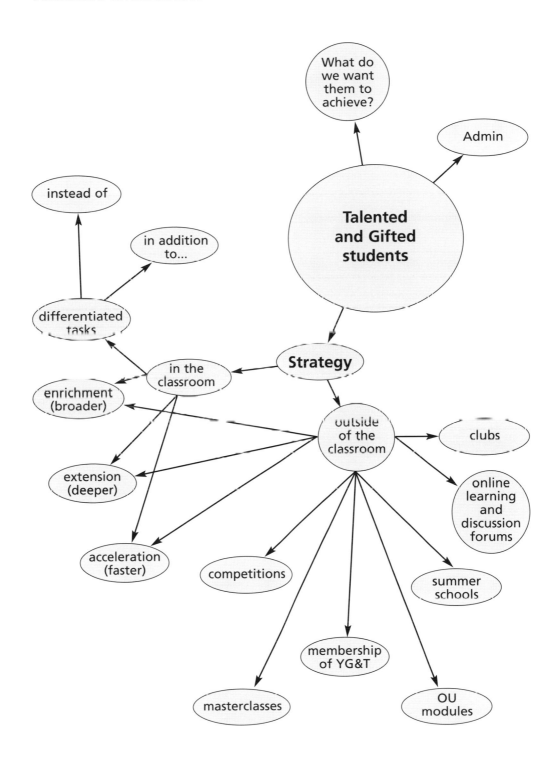

The aim of provision for gifted students

The focus here is on educational aims in a narrow sense. There are many possible emotional, social and spiritual benefits for giving a very able child appropriate learning experiences. These underpin the ethos of gifted provision but are omitted because of the risk of 'preaching to the converted'.

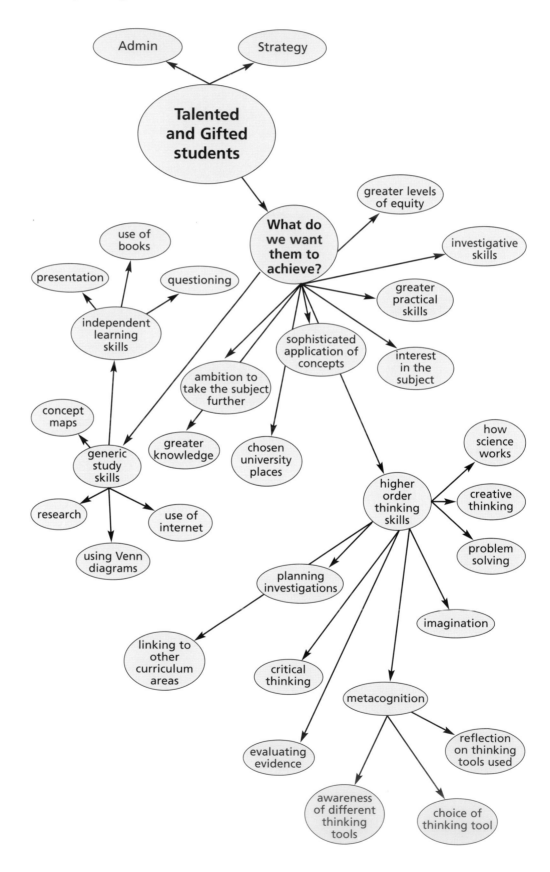

What we should do with gifted students

Materials are available to help whole science departments consider the issues around provision for gifted students at
www.standards.dfes.gov.uk/keystage3/downloads/agt40_mod4_sci.pdf
(accessed April 2007).

If giftedness is 'expertise in development' then we need to pause and ask what expertise in our subject is like. We need a vision for what high levels of performance means in chemistry. That vision depends on the developmental stage of the student. What we should aim to achieve with a student who is finishing their chemistry education at age 16 differs from the aims we may have for a student who follows a post-16 chemistry course.

Aspects of a vision for expertise in chemistry which are common with other subject areas:
- good independent learning skills;
- the ability to make links to other knowledge and concepts;
- communication skills;
- teamwork skills;
- generic study skills such as revision techniques, use of concept maps, Venn diagrams, flowcharts and other tools for ordering thoughts;
- higher order thinking skills – creative and critical thinking;
- metacognition – an awareness of the different thought processes that students use; and
- problem solving skills.

Aspects of expertise in students more specific to chemistry (or science):
- spatial awareness that enables students to visualise, in three dimensions, models of molecules, atoms and ions;
- appreciation of the nature of models as used in chemistry and an ability to choose the appropriate model for the situation in question;
- application of abstract models;
- use of published data, such as thermodynamic data, and awareness of its limitations;
- evaluation of the usefulness and reliability of models and data;
- practical skills;
- investigative skills;
- breadth and depth of knowledge of chemistry;
- ability to apply the fundamental concepts of chemistry to novel situations; and
- mathematical skills applied in a chemical context.

For post-16 students on a chemistry course there may be *fundamental* concepts that are not currently taught as part of the specification they are following. It is clearly detrimental for gifted students to miss out on these and they should be encouraged to find out about them and use them. Principally: *curly arrow mechanisms, entropy* and *the second law of thermodynamics* but others might want to add topics such as *molecular orbitals* to that list.

How the use of this resource fits in with the wider provision for gifted students

The model of gifted provision adopted nationally in England has a pyramid like structure, with classroom provision as a foundation to everything else. The bolt-on experiences outside the classroom should not be seen as sufficient provision for gifted students on their own.

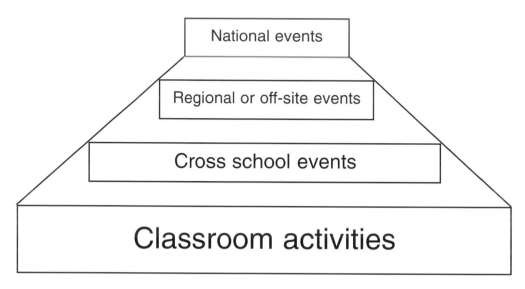

Figure 1 Structure of gifted and talented provision

The main purpose of this publication is to provide resources and ideas for *classroom activities* that might be used as differentiated tasks or whole class activities. Some of the teachers using this resource may be involved with chemistry provision for the very able beyond lessons, so it is worth considering what some of the opportunities might be outside of the classroom. Some of these are listed towards the end of this chapter under the heading *Opportunities for gifted students*.

Strategies

In general, provision for gifted students uses three approaches: enrichment, extension and acceleration. These are also referred to as broader, deeper and faster.

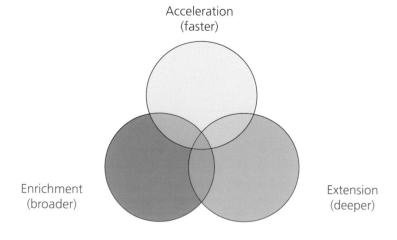

Figure 2 Three types of gifted provision

Enrichment, acceleration and extension
Enrichment activities are those that are outside the normal curriculum. Thus the learning experience is broadened beyond the scope of the standard provision.

Examples of ways in which curriculum topics can be enriched:

Contextual enrichment – approaching a topic via an application or situation, *eg* corrosion on the Titanic rather than just rusting.
Historical enrichment – studying the development of ideas, how models are changed and refined, *eg* role play around Priestly and Lavoisier as part of a topic on combustion.

Enrichment activities do not need to closely relate to the general curriculum. For example the activity *Norbert Rillieux and the sugar industry* in the RSC publication D. Warren, *Chemists in a social and historical context*, London: Royal Society of Chemistry, 2001.

Acceleration is covering concepts or knowledge earlier than would otherwise be done. Careful consideration needs to be given to acceleration as a strategy. You do not want to cause subsequent frustration by teaching the most able something which they are then going to be taught again.

Extension means taking a deeper look at current topics, using higher order thinking skills and applying theory to understand concepts to a greater depth than would otherwise be done. For example, relating chromatography to the particle model, planning experiments instead of following a recipe and evaluating different explanations of the same observation. A topic could be extended by approaching it via an egg race or problem solving scenario.

Many activities incorporate aspects of more than one approach as indicated in *Figure 2*.

The rationale for these resources

These resources focus mainly on extension activities. This was done partly to complement the scope of existing resources. Many of the materials already produced by the RSC and others offer opportunities to enrich the curriculum. Some suggested resources are listed later in this chapter under the heading *Other resources suitable for gifted students*.

Many of these resources are designed to be an episode that could fit into a lesson on a regular topic. The resources are designed so that they can be used to differentiate for the more able in a class or, if appropriate, for whole class use. Differentiation can put extra demands on teachers' time in class so for every activity there is a discussion of the answers and issues raised that are appropriate for the students. The *Discussion of answers* sheets are written **for the students** and can be given to them to review their own work or as guidance to help peer review and discussion.

Some of the activities are designed to develop study skills in a chemical context – *eg Organising your thoughts* and *Atoms, elements, molecules, compounds and mixtures*.

Some of the activities are designed to develop critical thinking skills – *eg Boiling point* and *Solutions*. Here the students are asked to evaluate a number of alternative explanations or ideas.

The *Covalent bonding* activity attempts to address the over reliance on 'filling the octet' as a model for bonding, as identified by a previous RSC Schoolteacher Fellow, Keith Taber.[4]

It is important that students develop an understanding of the nature of models in science. The activities *Bonding models* and *Formal charge* give them the opportunity to evaluate and refine models used in chemistry.

The concept cartoons – *eg Candle investigation* and *Ionic bonding* – are an easily reproducible way of adding a more challenging episode in a lesson. By offering new ways of looking at a situation they make it problematic and provide a stimulus for developing ideas further.[5] They can be used to develop critical thinking (evaluate several alternatives), creative thinking (produce several alternatives) and enquiry (asking several questions). They are easy for teachers to produce themselves and have the great advantage that there is not too much to read before you can start thinking.

Lateral thinking skills are important. *Rust* introduces lateral thinking problems to the students, where they search for the solution by asking questions that can be answered 'yes' or 'no' by the teacher.

Chemistry at its best is an interrelated web of concepts, practical skills, models and facts – each supporting and gaining support from the whole. Gifted students particularly appreciate a sense of overview of the subject and some of the activities in this resource – *eg Rates and equilibria* and *Entropy and equilibrium* – are written to show the connection between subject areas.

Students are often aware that examination questions have clear cut definite answers which the question, if read carefully enough, will point them towards. They are expected to reproduce well rehearsed thoughts and arguments. Examination questions ask students to state, account for, explain and describe. We do not often ask them to speculate about situations where the possible solutions are not well rehearsed and may not be known by the questioner or indeed at all. The activities *A new kind of alchemy* and *A question of thinking* encourage this type of speculation. Speculation involves *creative thinking skills* and helps to dispel the myth that everything interesting or exciting in chemistry is already known and understood.

Many students (especially gifted ones), when faced with a question, are able to leap to an answer without having to explore their thinking about it (as illustrated by the top part of *Figure 3*). The result is sometimes that they do not have the explorative skills to consider a problem if they do not see an answer immediately and then tend to declare 'I don't get it'. Toddlers will sometime pick up objects and explore them. They will turn them over in their hands and scrutinise them from every angle, try tasting and hitting them and establish whether they will fit in their nostrils *etc*. I believe that students need identified thinking tools to help them explore questions in science.

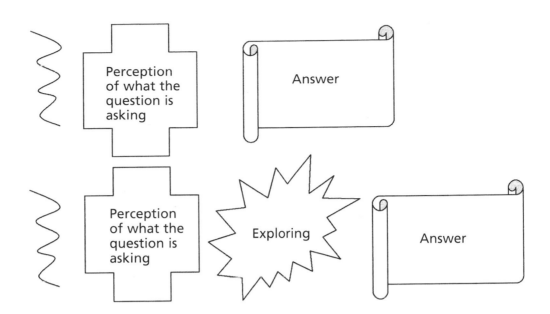

Figure 3 A diagram showing the place for explorative thinking skills
in problem solving

Part of the answer lies in the students developing **metacognition** (thinking about thinking) skills. There is a good case for frequently discussing with students what kind of thinking strategies they employed in answering questions and identifying specific thinking procedures they can use when they meet a question where the answer is not readily apparent. Instead of supplying the answer in response to the declaration 'I don't get it', it is better to discuss what thinking procedures could be employed when searching for understanding

Some thinking procedures might be:

- organising the question into what we know and what we need to find out;
- rewording the information given;
- listing all the facts and concepts that might have a bearing on the question;
- developing a flowchart from the information given – *eg* information A tells us that trend B will be true *etc.*
- developing a flowchart back from the answer – *eg* to be able to give a value to Z I will need to know Y which will be determined by X;
- brainstorming alternative suggestions;
- 'mind mapping' out from the original problem;
- fishing for a similarity with a familiar fact or concept in chemistry;
- briefly exploring, or identifying an intuitive feel for, which direction a brainstormed suggestion might take;
- evaluating alternatives and prioritising the most likely suggestions;
- having arrived at an answer go back and ask if it seems reasonable; and
- acting as a critic of your solution. Are there identifiable problems (sources of error, dubious assumptions *etc*) with the solution?

The activities *A question of thinking*, *Mixing drinks* and *Volume changes* are designed to develop students' metacognition.

The Chemistry Olympiad questions, which have been selected from the RSC Round 1 selection papers for the *International Chemistry Olympiad*, are carefully written to test and develop the problem solving skills of the most able students. They often introduce an interesting context in which the chemistry is explored. By providing a more detailed discussion of the answers it will hopefully encourage teachers to give these out as part of the work on the relevant topic without fear that it will demand lots of time from them to explain or work out the solutions. More students will hopefully feel encouraged to sit the Olympiad Round 1 paper having tackled past paper questions.

Some post-16 chemistry courses currently leave out important topics such as curly arrow mechanisms, entropy and the second law of thermodynamics. Several activities refer to these topics and *Curly arrows and stereoselectivity in organic reactions* and *The second law of thermodynamics* are designed to develop students' understanding of these topics.

Many gifted students have good memories and recall facts with little effort. However, they do not all recall facts well and some may have a tendency to categorise factual information as trivial and therefore not worth committing to memory. The puzzles and Su Doku activities encourage students to engage with factual information in a problem solving context. All kinds of puzzles – *eg* crosswords and Su Doku – are attractive, almost addictive, to some, and can be a more exciting approach than being told to learn facts for a test. The logical problem solving skills developed are useful in themselves. For example, they are similar to the skills used to assign NMR spectra or identify an unknown.

How to use these resources

As many of these activities are designed to add a more challenging episode to regular topics, they could be given to students who finish the general tasks early. However, some students can be discouraged by the perception that the reward for staying on task is yet more work. A preferred option is to use these activities **instead** of some less challenging work for those students who will not be sufficiently challenged by the alternative task. Teachers may wish to select which students do the extension activities on the basis of general ability, recent good work, or the students demonstrating that they have a good grasp of the topic already. It is often beneficial for teachers to find out at the start of a topic what the students already know; students who demonstrate a sufficient knowledge or understanding at the start could then be asked to take on the extension activity. Another model for consideration is **self selection** – ask the students to make a judgement as to how well they understand a topic and to select the level of challenge that they believe they are ready for.

Many gifted students learn well and are most creative when they discuss ideas. It is not always necessary for their answers to be written down and there is a positive benefit in offering a variety of methods of reporting back their thoughts. Many of the activities ask several questions which can discourage students if they are expected to produce written answers for all of them. If there is more than one student working on an activity it is recommended that at least some of the questions are simply discussed.

Assessment of the students' responses can be done at several levels. There is often a good case for peer review, particularly among gifted students. This gives them some feedback and also develops their listening, empathy and communication skills. They should be encouraged to read through the *Discussion of answers* and may indeed decide that they disagree with them. The activities may well stimulate questions that they will want to discuss with their teacher.

Some personal reflections by the author

Over the course of the fellowship year I have read and thought about the teaching and learning of students (gifted and otherwise) to a greater degree than I had done previously. I go back to teaching with several resolutions about how my teaching can be improved and I record them here in the hope that they will be helpful for other teachers.

- Avoid 'experiments to show...' where the student can already predict the outcome.
- Set more open ended tasks and encourage a greater diversity of reporting methods.
- Do more investigation work (with a sigh of relief that it is not coursework) where the students (and teacher) do not know what the outcome is.
- Allow gifted students to learn more by discussion and less by writing.
- Repeatedly ask students if there are other additional possible solutions when they have arrived at an answer.
- Foster enquiry by asking what questions could be asked.
- Adopt a must, should, could model in the scheme of work which gives examples of relevant extension and enrichment activities.
- Use more of the problem solving activities in *In search of solutions* *www.chemsoc.org/networks/learnnet/solutions.htm* and *In search of more solutions* *www.chemsoc.org/networks/learnnet/more_solutions.htm*.

'Often there has been too much reliance by the teacher upon absolute right answers related to some sense of truth, or sometimes too much control on the student's thinking within a set paradigm.'[6] Appreciate whatever thinking or logic might lie behind an answer even when the answer is wrong.

Edward de Bono[7] divides thinking into 'reactive' and 'pro-active', where reactive thinking is in response to a defined problem but pro-active thinking is more open ended, creative and diverse. I would like to give students greater opportunity for pro-active thinking. I hope to risk setting some very open ended tasks such as: producing presentations on an application of chemistry; coming up with fundamental questions they think would have important answers; describing a day in the life of someone in a world suddenly deprived of man-made materials; predicting how chemistry will transform our lifestyles in 40 years time; describing the biochemistry of life evolved on a planet with as much sulfuric acid or bromine as we have water *etc*.

Opportunities for gifted students

Young Gifted & Talented (YG&T)

Membership of YG&T gives access to a wide range of provision for gifted and talented students aged 4–19, across all subject areas. Details are available at *www.dcsf.gov.uk/ygt* (accessed August 2007).

Chemistry clubs

The Salters' Institute of Industrial Chemistry has produced two handbooks full of ideas for mainly practical activities for use in chemistry clubs. These are available free on CDROM or to view on-line. Details are available at *www.salters.co.uk/club/publications.htm* (accessed April 2007).

Off-site visits

Opportunities for these vary depending on location but there is a strong feeling that students gain a great deal from 'meeting an expert'. Details of *Chemistry at Work* events can be found at *www.rsc.org/Education/chemwork/index.asp* (accessed July 2007).

Competitions

There are descriptions of several national competitions run by the RSC and the Salters' Institute on the web. These include *Top of the Bench* (14–16 years old), the *Schools Analyst* competition (lower sixth form or equivalent), the *Chemistry Olympiad* (post-16) and the *RSC Bill Bryson Prize*. For details visit *www.rsc.org/events* (accessed July 2007).

Year 7–9 (England and Wales), P7–S2 (Scotland), Year 8–10 (Northern Ireland), Primary Year 8–Secondary Year 2 (Republic of Ireland)

The quarterly publication *A Pinch of Salt* from the Salters' Institute contains a paper competition with entries to be posted by a certain date. Your school may receive *A Pinch of Salt* in the post and the publication can be downloaded at *www.salters.co.uk/institute/publications.htm* (accessed April 2007).

Year 7 or 8 (England and Wales), P7 or S1 (Scotland), Year 8 or 9 (Northern Ireland), Primary Year 8 or Secondary Year 1 (Republic of Ireland)

Salters' festivals (in association with the RSC) are a one day competition, involving practical and written tasks for teams of four. No preparation is necessary and in the past there has been help to finance travel. The *festivals* are hosted by numerous universities across the UK. For details visit *www.salters.co.uk/festivals/* (accessed April 2007).

Year 10 (England and Wales), S3 (Scotland), Year 11 (Northern Ireland), Secondary Year 3 (Republic of Ireland)

Chemistry Camps are run by the Salter's Institute on behalf of the Chemical Education Group, which includes the RSC, in several universities across the UK. The camps last four days and three nights and are hugely popular with the students who attend. For more details visit *www.salters.co.uk/camps/index.htm* (accessed April 2007).

Year 9–11(England and Wales), S2–4 (Scotland), Year 10–12 (Northern Ireland), Secondary Year 2–4 (Republic of Ireland)

Top of the Bench is a team competition organised by the RSC, which has a regional round

RSC | Advancing the Chemical Sciences

followed by a national final held in the Science Museum in London. For more details visit *www.rsc.org/TOB* (accessed April 2007).

Sixth form (or equivalent)

The RSC Analytical Division *Schools' Analyst* competition is a practical based competition organised by the RSC. For more details visit
www.rsc.org/Membership/Networking/InterestGroups/Analytical/SchoolsComp/index.asp (accessed April 2007).

The *Chemistry Olympiad* is for students following post-16 chemistry courses. There is an online test for lower sixth (or equivalent). For details visit
www.chemsoc.org/networks/learnnet/olympiad_L6.htm (accessed April 2007).
The Round 1 paper test is aimed at upper sixth form (or equivalent) students, although lower sixth (or equivalent) students can also take part. It contains carefully constructed questions that test both chemical understanding and problem solving skills. Certificates and prizes are awarded on the basis of Round 1 results and those who scored exceptionally well go on to a selection procedure for the UK team to take part in the *International Chemistry Olympiad www.rsc.org/olympiad* (accessed July 2007).

Independent learning projects

The Open University have designed a set of courses for post-16 students as part of their *Young Applicants in Schools Scheme* (YASS). Each course represents approximately 100 hours of study and can be completed over the summer holidays. Students gain 10 points towards an OU degree per course completed. None of the courses currently available cover a lot of chemistry but they greatly develop independent learning skills. The course which has the most overlap with chemistry is *Understanding human nutrition* (code SK183), but there is a range of other titles that might be of interest to particular students. There is a charge for these but some students may be able to obtain help with meeting the cost. Further information can be found at *www.open.ac.uk/science/short* (accessed April 2007).

Other resources suitable for gifted students

1. Approaching topics in a problem solving context is beneficial to students of all abilities but has a particularly motivating effect for the most able. The RSC resource *In search of solutions* is available free to download at
   www.chemsoc.org/networks/learnnet/solutions.htm.
   A description of *In search of more solutions* is available at
   www.chemsoc.org/networks/learnnet/more_solutions.htm.
   Both are particularly well suited to able students.

   C. Wood, *Creative problem solving in Chemistry*, London: Royal Society of Chemistry, 1993 is a book of 30 group activities that develop higher order thinking skills.

2. Some post-16 students have a curiosity that goes beyond the limitations of what teachers are able to cover in class. A book ideal for the most able post-16 students to read from cover to cover is J. Keeler and P. Wothers, *Why chemical reactions happen*,

Oxford: Oxford University Press, 2003. The fundamental principles behind chemistry are explained to a greater depth than in post-16 specifications. Copies of this book were distributed to schools and colleges with sixth forms in spring 2006 by the RSC.

3. Some RSC resources useful for contextual enrichment are:
   Contemporary chemistry for schools and colleges
   Inspirational chemistry
   Kitchen chemistry
   Industrial chemistry case studies
   The nature of science
   Climate change
   Health, safety and risk

   Details of these are available at
   www.chemsoc.org/networks/learnnet/resources.htm.

4. Other useful context enrichment resources:
   Hill and Holman, *Chemistry in context: Laboratory manual and study guide*,
   Cheltenham: Nelson Thornes Ltd, 2000.
   Jeffrey Hancock, *Sand castles and mud huts* (now out of print but you may have a copy somewhere in school).

   Infochem, the student supplement in *Education in Chemistry*, has contextual information as well as competitions and other features. Permission is granted for the material to be copied for use in schools.

5. *The Gatsby Science Enhancement Programme (SEP).*
   In particular *Enriching school science for the gifted learner*, a publication designed to support science teachers responsible for addressing the needs of the most able students at KS4.
   Enriching school science for the gifted learner, London: Gatsby Educational Projects, 2007.

6. References
   [1] Young Gifted and Talented Youth (YG&T) statement of beliefs.
   [2] B. Teare, *Effective provision for able and talented Children*, Stafford: Network Educational Press, 1997.
   [3] D. Warren, *Chemists in a social and historical context*, London: Royal Society of Chemistry, 2001.
   [4] K. Taber, *Chemical misconceptions – prevention, diagnosis and cure*, London: Royal Society of Chemistry, 2002
   [5] S. Naylor and B. Keogh, *Concept cartoons in science education*, Sandbach: Millgate House, 2000.
   [6] P. O'Brien, *Using science to develop thinking skills at Key Stage 3*, London: David Fulton, 2003.
   [7] E. de Bono, *Teach your child how to think*, London: Penguin, 1992.

RSC | Advancing the Chemical Sciences

How to use this resource

Nearly all the activities in this resource can be used as differentiated activities for a selection of more able students in a mixed ability group. The aim is that the students should be able to work through the activities without needing too much teacher support. So that the students can get feedback on their work for each activity there is a *Discussion of answers* sheet which is **for the students**. This can be used for students to check their own work or to help them review the work of other students.

Most of the activities relate to a particular curriculum area which, in most cases, is clear from the title. The activities can then be done when the class is doing the relevant topic. Several of the activities could serve as more challenging alternatives to episodes in the lesson which would be less appropriate for the most able students.

The teachers' notes and introduction are included in the book and the *Student worksheets* and *Discussion of answers* are available on the CDROM.

CDROM instructions and system requirements

The CDROM is fully compatible with Windows® NT/2000/XP/Vista and can be used on most other computer systems equipped with a CDROM drive.

In addition you will need:

- Web browser – the content has been optimised for Internet Explorer® 6 but will function correctly using most other browsers. An Internet connection is not required.
- Java – to use the advanced search facilities Java must be installed and enabled.
- Adobe Acrobat® PDF reader – required to open PDF resource files.
- Microsoft Excel® – required to open Excel® resource files.
- Microsoft PowerPoint® – required to open PowerPoint® resource files.

To use, insert the CDROM into the CDROM drive.

PC users: your PC should run the CDROM automatically. If it does not, open the CDROM using My Computer and run the programme GandT. You may access the resources directly from the CDROM or else install them to your PC's hard disk. Alternatively, use your web browser to open the file index.htm.

Users of other computer systems: using your web browser, navigate to the CDROM and open the file index.htm.

The CDROM licence allows the files on the CDROM to be downloaded and to be accessible over a network. The RSC will not offer support or guidance on how best to network the files.

Note about fonts

If, when opening any of the files on the CDROM, you find certain chemical symbols to be incorrectly represented (*eg* the equilibrium symbol), then please download the *Royal Society of Chemistry font* from *www.chemsoc.org/networks/learnnet/rscfont.htm*. The website page carries full instructions for the simple procedures involved in downloading the RSC font and using it in Microsoft® applications such as Word®, PowerPoint® and Excel®.

Note about printing the PDFs

If you encounter a problem where the student sheets print out slightly smaller on the page than you expected, make sure that in earlier versions of Adobe Acrobat, the option for 'Fit to page' on the print dialogue box is unchecked. In more recent versions of Adobe Acrobat® this option is found by first choosing the 'Properties' button in the print dialogue box and then the 'Effects' tab. Choose 'Actual size' not 'Fit to page'.

Disclaimer

The CDROM has been thoroughly checked for errors and viruses. The RSC cannot accept liability for any damage to your computer system or data which occurs while using this CDROM or the software contained on it. If you do not agree with these conditions, you should not use the CDROM.

Atoms, elements, molecules, compounds and mixtures

Student worksheet: CDROM index 01SW

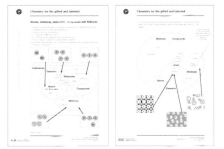

Discussion of answers: CDROM index 01DA

Topics

Atoms, elements, molecules, compounds, mixtures, structure and different representations of bonds and atoms.

Level

More able 11–13 year old students.

Prior knowledge

The differences between elements, mixtures and compounds.

Rationale

This activity is designed to help students clarify the relationship between various parts of their knowledge in these topics and develop the skill of using Venn diagrams in organising their understanding.

Venn diagrams are a method of organising your thoughts like lists or mind maps. The particular advantage of Venn diagrams is that they clearly show the interconnection or overlap between categories. For example, a substance may be molecular and an element, or molecular and a compound. A reaction may be feasible and proceeds at room temperature, or feasible but not occur because of high activation energy. They also show where characteristics are mutually exclusive: this substance, if a compound, cannot be an element.

Different representations of atoms and molecules are used to show that there are several different ways that they can be visualised.

Use

The activity could be used as an extension activity for students at any stage after the main topics listed have been taught. It could be used as a follow up to a lesson on compounds and mixtures.

To extend the ideas here, students could be asked to think of other substances to be added to the diagram or to design a new Venn diagram for elements (with subsets: metals, non-metals, metalloids, gases, solids, liquids *etc*).

Boiling point

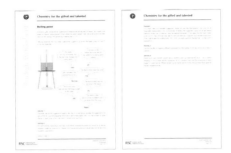

Student worksheet: CDROM index 02SW

Discussion of answers: CDROM index 02DA

Topics
Boiling points and measurement and errors. The link between boiling point and atmospheric pressure is also discussed in the *Discussion of answers* sheet.

Level
Middle to high ability students in the 11–13 age range.

Prior knowledge
The boiling point of water.

Rationale
This activity is designed to help students develop their critical thinking (evaluative) skills. They are presented with a surprising measurement for the boiling point of water and several suggestions as to why it might be. The main task is to evaluate the different suggestions. The experimental nature of science is reinforced by asking them how each suggestion could be tested practically in Activity 2, which could be approached as a class discussion. Creative thinking skills are used in Activity 3 when students add suggestions of their own.

In Activity 4 they are asked to devise a concept cartoon. One scenario is presented with diagrams and as few words as possible – alternative explanations, suggestions or questions are presented in speech bubbles. Any good concept cartoons generated could be tried out on other students to evaluate the alternative suggestions. Good examples of concept cartoons to show the students can be found in *Concept cartoons in science education*[1] or from other activities in this publication such as *Candle investigation*. A good idea might be to do a group effort on one together on the board before the students produce their own.

Use

This activity is best used as a whole class task. Although it was written with more able students in mind it is accessible for most abilities with support. It could be used when boiling points are taught or after an experiment where qualitative data are recorded as it facilitates a discussion on the reliability of data.

When the students have completed the worksheet they should be given the *Discussion of answers* sheet. They could check their own work or conduct a peer review of the work of another student or group.

[1] S. Naylor and B. Keogh, *Concept cartoons in science education*, Sandbach: Millgate House, 2000.

Candle investigation

Student worksheet: CDROM index 03SW

This worksheet is also provided on the CDROM as both a PowerPoint® presentation *03PP* and a teacher-editable Word® document *03WD* – please refer to the *Use* section below.

Discussion of answers: CDROM index 03DA

Topics

Combustion and reliability of data.

Level

Able students aged 11–13.

Prior knowledge

Combustion requires oxygen.

Rationale

Experiments 'to show' can be frustrating for very able students if they already know what the experiment will show. The approach used here is to change the experiment into one for which they will not know what happens.

They have to use critical thinking to evaluate the alternative suggestions in the concept cartoon. They can evaluate some real experimental data… this might stimulate motivation to carry out the experiment to obtain better evidence. The students are then asked to briefly think about the difficulties of showing slight effects, as in medical research.

They need to use creative thinking to produce their own concept cartoons. Good examples of concept cartoons to show the students can be found in *Concept cartoons in science education*[1].

Use

This could be used to follow on from a class investigation into the effect of beaker size on the length of time the candle burnt. The PowerPoint® version (CDROM index 03PP) could be used as a focus for class discussion, using a projector as an alternative to the worksheets.

The Word® version (CDROM index 03WD) could be used as a differentiation activity for more able students in the group who can accurately predict the outcome of the size of beaker experiment.

When this activity was trialled, the students enjoyed making their own concept cartoons. These could be tried out on other students and a prize could be offered for the best one.

If the students plan and carry out their own investigations then it will be an activity for a whole lesson, otherwise it could simply be an episode at the start or end of a lesson.

[1] S. Naylor and B. Keogh, *Concept cartoons in science education*, Sandbach: Millgate House, 2000.

Chromatography

Student worksheet: CDROM index 04SW

Discussion of answers: CDROM index 04DA

Topics

Chromatography, modelling and particles.

Level

Able students aged 11–13.

Prior knowledge

The students should have practical experience of paper chromatography and particle theory.

Rationale

This activity extends the students' understanding of chromatography. It links chromatography with particle theory and develops the tools of analogy and modelling.

Use

This activity could be used as extension work that the more able in a group can do individually or it could be done as a whole group. If a whole group of students is doing the activity they could develop their ideas about Question 1 into a concept cartoon[1]. Students could pool their results of the modelling experiment to get a greater number of steps.

[1] S. Naylor and B. Keogh, *Concept cartoons in science education*, Sandbach: Millgate House, 2000.

This page has been left blank intentionally.

Universal indicators

Student worksheet: CDROM index 05SW

Discussion of answers: CDROM index 05DA

Technician's notes: CDROM index 05TN

Topics

Indicators, universal indicators and pH.

Level

Able students aged 11–13.

Timing

The full worksheet should take a double period. There is quite a bit of flexibility in how many binary mixtures of indicators they try.

Prior knowledge

Acids, alkalis and indicators.

Rationale

This activity develops understanding of universal indicators and single indicators. The students build up their understanding by mixing two indicators. They also develop an awareness that the observed colour may be due to a mixture of colours.

Use

This activity is best used as an extension to work on single indicators and pH.

Alternative

An alternative activity with a less structured problem solving approach would be *Making your own indicator* Activity 1 in the RSC publication *In search of solutions* which can be downloaded at ***www.chemsoc.org/networks/learnnet/solutions.htm*** (accessed April 2007).

Apparatus (per group)

Eye protection

At least four test-tubes

A test-tube rack

Dropping pipettes

Chemicals (per group)

50-100 cm^3 of pH 3 buffer solution

(An alternative is 0.5 mol dm^{-3} ethanoic acid)

50-100 cm^3 of pH 4 buffer solution

50-100 cm^3 of pH 7 buffer solution

50-100 cm^3 of pH 10 buffer solution

Methyl red indicator solution

Methyl orange indicator solution

Bromothymol blue indicator solution

Phenolphthalein indicator solution **(Highly flammable)**

Safety

It is the responsibility of the class teacher to consult an employer's risk assessment for this experiment.

Which is the odd one out?

Student worksheet: CDROM index 06SW

Discussion of answers: CDROM index 06DA

Topics

Metals, alloys, acids and alkalis, particles, apparatus, elements, the Periodic Table, types of reaction, fixed points and separation techniques.

Level

Students in the 11–13 age range, middle to high ability.

Prior knowledge

A general awareness of chemistry taught to the 11–13 age range.

Rationale

This activity is fairly synoptic in nature and should encourage a rapid consideration of the range of concepts met in chemistry taught to ages 11–13. As scientists we often survey the available models or concepts to decide which is most pertinent to the current problem. This activity is designed to develop those skills. The students will need to think laterally in some cases. It may help students develop the skills needed for synoptic exams where the questions could be about a number of topics.

Use

The activity is synoptic and could be used as a stimulus for revision towards the end of a course. The students should be asked to work through the questions in groups of two or three. The answers could be discussed as a whole group or the *Discussion of answers* sheet could be given to the students.

The students should then be encouraged to devise their own questions and answers (perhaps for homework) which could be tested on others in their group.

This page has been left blank intentionally.

Rusting

A Titanic Rust Problem

Class discussion (teacher-led) from a PowerPoint® presentation: CDROM index 07CD

There is **no** student worksheet for this activity.

Discussion of answers: CDROM index 07DA

This material is also provided on the CDROM as both a PowerPoint® presentation *07PP* and a teacher-editable Word® document *07WD* — please refer to the *Use* section overleaf.

Technician's notes: CDROM index 07TN

Topics

Rusting, corrosion and sacrificial protection.

Level

Able students aged 11–13.

Prior knowledge

Rusting of iron requires oxygen and water. Salt speeds up rusting. The sacrificial protection of iron by more reactive metals, such as zinc or magnesium.

Rationale

This activity looks at rusting in the context of shipwrecks. It aims to develop higher order thinking skills including some lateral thinking and creative thinking. It has different demands to the traditional experiment to show the factors needed for rusting to occur.

Use

It is best used as a teacher-led class discussion using the PowerPoint® presentation (07CD), leading to group and practical work. The first parts could be used as an episode to follow on from an introduction to rusting or for students who already know which factors are required for rusting. The planning exercise at the end is optional and the activity can be used without it.

It starts with a lateral thinking exercise. The students may not have met these before and you might want to go through an example with them – 'Anthony and Cleopatra lie dead on the floor (in a pool of water)'. The students ask questions and in the end discover that Anthony and Cleopatra are goldfish whose bowl has fallen on the floor when the shelf, on which it was sitting, broke.

In the lateral thinking exercise in the activity the shipwreck in deeper water was carrying a cargo of zinc and magnesium. Gifted students could be asked to devise their own lateral thinking problems, perhaps for homework.

The students are asked to design a concept cartoon on the predictions about the shipwrecks. This can be done all together or in groups. It should have a drawing in the middle showing the shipwrecks at different depths and four speech bubbles around it expressing opinions about why one will be rustier than another. The best cartoons will be those with the greatest number of plausible explanations. Good examples of concept cartoons to show the students can be found in *Concept cartoons in science education*[1] or from other activities in this publication such as *Candle investigation*. A good idea might be to do a group effort on a different topic on the board before the students produce their own. One example could be four opinions on: whether mayonnaise is a liquid or a solid; whether magnesium would dissolve in acid on a space station; what climate change will mean for the UK; who is doing the greater good for humanity, the doctor or the research scientist working on a cure for cancer.

If you want the students to carry out their planned investigations it will need some preparation in advance. A suggested list of some apparatus required is included for technicians.

The students may find getting reliable results difficult and should be encouraged to run double experiments so that they get an idea of the reliability. A couple of methods they could try for measuring the amount of rusting are weighing the dried rust after filtering or weighing the nails and recording mass lost as rust.

The *Discussion of answers* is also given in Word® (07WD) and PowerPoint® (07PP) formats to suit the teacher's requirements. Selected slides from the *Discussion of answers* sheet could be incorporated in the main PowerPoint® presentation (07CD) to avoid jumping from one to the other.

[1]   S. Naylor and B. Keogh, *Concept cartoons in science education*, Sandbach: Millgate House, 2000.

Solutions

Student worksheet: CDROM index 08SW

Discussion of answers: CDROM index 08DA

Topics

Dissolving, particles and hydrated crystals.

Level

Able students in the 11–13 age range.

Prior knowledge

The particulate nature of matter.

Rationale

This activity is designed to develop the students' higher order thinking – particularly critical thinking skills – in the context of solutions. The students have to apply some particle theory and logical reasoning to simple experimental observations.

Use

This could be given to a whole class or part of a class as part of more general work on dissolving and solutions. The students may need encouragement to think in terms of reasoning and logic rather than simply reciting facts.

When the students have completed the worksheet they should be given the *Discussion of answers* sheet. They could check their own work or conduct a peer review of the work of another student or group.

This page has been left blank intentionally.

Astounding numbers

Student worksheet: CDROM index 09SW

Discussion of answers: CDROM index 09DA

Topics

Avogadro's number, the charge on an electron and proton, the density of a nucleus and the mass deficit of a helium nucleus.

Level

Very able pre-16 students or able post-16 students.

Prior knowledge

Students will need to be able to use standard form and rearrange equations.

Faraday's constant is quoted in Coulombs; students who have not met units of charge previously could leave out question 10.

Rationale

This activity is designed to be fun and generate a wow factor for the students. It tries to convey how amazing the scale is that we use in chemistry. In question 8 the students are asked to use creative thinking and come up with some questions of their own.

The students are asked to estimate some of their answers and they should appreciate that it may be desirable to work with fewer significant figures in these cases.

Use

This could be given to the more able students in a class, who have mastered mole calculations while the others are still practising.

More than one student needs to do the activity for them to try out their questions on each other. To finish off they could be asked to give a short presentation to the rest of the group along the lines of 'Did you know 1 mole of...'

There are some hints they can be given if they run out of ideas for question 8. Where a quantity needs to be estimated the students could be asked to try an internet search or simply guess – *eg* the number of grains of sand in 1 cm^3 could be estimated and the estimate written into the question in the same way that the mass of a double-decker bus is in question 9.

How big is the beach with a mole of sand grains?
How big is an omelette made with a mole of eggs?
How far could a snail crawl in a mole of seconds?
How big is an ocean with a mole of buckets?
How deep is the flood with a mole of raindrops?
How many men would it take to produce a mole of sperm?
How large is a forest with a mole of leaves?

The answer to question 12 in the *Discussion of answers* sheet

The answers from the two methods are different. Which one do you think is a better estimate? Why?

I slightly favour the second method (answer 4.7 million) because it does not assume that there is no empty space in the water. However, it does assume a regular structure and it may have been better to use the density of ice rather than of water to base the calculation on. The density of ice is approx 0.92 g cm^{-3} and gives a final answer, using method two, of 4.8 million.

Bonding models

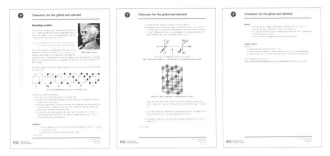

Student worksheet: CDROM index 10SW

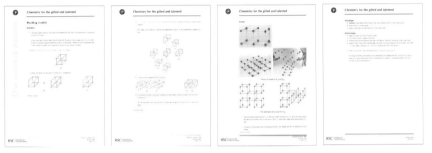

Discussion of answers: CDROM index 10DA

Topics
How models are used in science, refining a model, geometry and coordination number and bond angles.

Level
Very able pre–16 or post–16 students.

Prior knowledge
Dot and cross diagrams and covalent and ionic bonding.

Rationale
Many students complain during their post-16 chemistry course that their teachers 'lied' to them in their pre–16 course. This is particularly true when revisiting atomic structure and bonding in post–16 courses. What the students are failing to appreciate (and perhaps we are failing to teach) is the nature of scientific models and how they are used in science. The general principle, that we use the simplest model available that works for the situation under consideration, escapes them. We refine or replace the model when it fails to explain or predict observed phenomena. A model should not be regarded as **truth** but as a useful systematic way of explaining or predicting events. Students may well be aware of an example of this in physics, that at low speeds Newtonian mechanics works fine but at speeds approaching the speed of light Einstein's special theory of relativity needs to be used.

Some students will develop the skill of holding alternative models together in their minds and choosing which to use based on the particular question – *eg* alternative models for the bonding in benzene or estimating the degree of ionic or covalent character in a bond by

Fajan's rules and electronegativity values.

This activity gets the students to think of the model they have been taught as a model rather than the truth.

Use

Teachers may need to explain to the students that the aim of the activity is to teach about the nature of models rather than giving them a new model of bonding which they will use afterwards. This could be used with a whole class or as a differentiated activity for part of a class.

Each group will need some Plasticine® (preferably two colours) and several cocktail sticks.

 This symbol means those questions are best tackled as a discussion if a group of students are doing this activity.

When the students have completed the worksheet they should be given the *Discussion of answers* sheet. They could check their own work or conduct a peer review of the work of another student or group.

Ionic bonding

Student worksheet: CDROM index 11SW

Discussion of answers. CDROM index 11DA

Topics

Ionic bonding, electrostatic attraction and energetic stability.

Level

Very able pre–16 students.

Prior knowledge

Atomic structure, ionic bonding and the formation of ions.

Rationale

This activity helps students to think through the importance of the electrostatic attraction between ions to the model of ionic bonding.

Use

This could be used as an extension activity for very able students after a lesson on ionic bonding.

The instructions can be given verbally or printed in the first sheet. Students should work in groups of two or three.

This page has been left blank intentionally.

Noble gases

Student worksheet: CDROM index 12SW

Discussion of answers: CDROM index 12DA

Topics
Trends in the noble gases, spotting patterns in data and the reactivity of noble gases.

Level
Very able pre–16 students.

Prior knowledge
Simple kinetic theory, atomic structure, trends in reactivity in Groups 1 and 7.
The mole is useful for question 4 but very able students should be able to do without it.

Rationale
The activity sets some critical thinking and pattern spotting tasks in the context of the noble gases. The students are given data that can be manipulated to show a directly proportional relationship, they may well develop their skills at determining mathematical relationships between variables from graphs.

Use
This can be used as a differentiated activity for a group of very able students who already know or will readily acquire what the specification requires them to recall about the noble gases. It could be used as a whole group exercise with support for question 4, where the students could be prompted to plot specific heat capacity against 1/RAM. They could be asked to consider the statements as hints towards the answer.

When the students have completed the worksheet they should be given the *Discussion of answers* sheet. They could check their own work or conduct a peer review of the work of another student or group.

Note

The specific heat capacity values quoted are measured at constant pressure. The average value for the noble gases is approximately $5/2$ R J K^{-1} mol^{-1}. C_v values (specific heat capacity measured at constant volume) would be approx $3/2$ R J K^{-1} mol^{-1} (for monoatomic gases) because no work is done against the external pressure as the temperature is raised because the gas is not allowed to expand.

Organising your thoughts

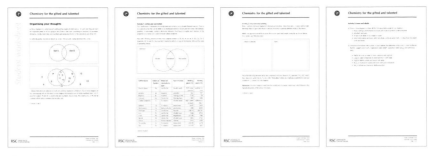

Student worksheet: CDROM index 13SW

Discussion of answers: CDROM index 13DA

Topics

Activity 1: metals, metalloids and non-metals; Activity 2: structure and bonding;
Activity 3: alkalis and bases.

Level

Able 14–16 year old students.

Prior knowledge

Activity 1: characteristics of metals; Activity 2: giant and molecular structures;
Activity 3: pH.

Rationale

These activities are designed to help students clarify the relationship between various parts
of their knowledge in these topics and develop the skill of using 'Venn'-like diagrams
in organising their understanding. The Venn diagram is introduced with a simple
mathematical example. Students then do three activities that progressively stretch their
skills in the context of the chemical topics of: elements, structure and bonding and, finally,
alkalis and bases.

Use

The worksheet was written for students to work through the introduction and all three
activities in one go, but each of the three activities could be done separately (with the
introduction if required) when those topics are most relevant in a student's course. Activity
2 suggests a concept map (or mind map/spider diagram) could be drawn flowing out from
a central Venn diagram. This could be set as homework and the students given a sheet of
A3 paper for adequate space to display their knowledge and understanding of the topic.

When the students have completed the worksheet they should be given the *Discussion of answers* sheet. They could check their own work or conduct a peer review of the work of another student or group.

Elemental Su Doku

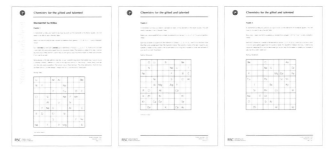

Student worksheet: CDROM index 14SW

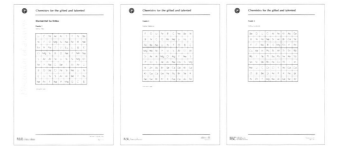

Discussion of answers: CDROM index 14DA

Topics
Groups and periods in the Periodic Table

Level
Middle to high ability students in the 11–16 age range.

Prior knowledge
The numbering of groups and periods in the Periodic Table.

Rationale
This activity helps students gain a familiarity with the Periodic Table by getting them to continually refer to it in a problem solving activity. Students should more easily recall the elements in the same group after doing some of these problems. The logical skills developed in solving the puzzle are useful.

Use
This activity could be used at any time but is best used in conjunction with work on the Periodic Table. Three elemental Su Doku puzzles are given with answers. Students should start with puzzle 1 as this contains the most detailed explanation of how to solve it.

Further puzzles can be readily generated using the Excel® templates in the post-16 sections *DIY Su Doku* CDROM index 35EX and *DIY ionic Su Doku* CDROM index 36EX. Instructions for doing this can be found on pages 71–74.

This page has been left blank intentionally.

Ionic Su Doku

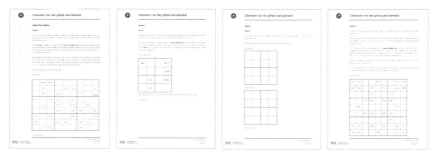

Student worksheet: CDROM index 15SW

Discussion of answers: CDROM index 15DA

Topics

Working out the charges on ions from formulae and working out formulae from the charges of ions.

Level

Middle to high ability students in the 14–16 age range and post-16 students.

Prior knowledge

How the charges of ions determine formulae of ionic compounds and the charges of some ions.

Rationale

This activity gives students practice at working out formulae in a problem solving context.

In puzzle 1, to avoid any confusion, the O^{2-} ion is assumed for all oxides including the Group 1 metals. Hydrogen is included as one of the cations so the students recall the formulae of the acids.

Puzzle 2 gives the students practice of working out the charges of ions from formulae.

Use

This activity could be used at any time but is best used in conjunction with work on ionic bonding and formulae. Three ionic Su Doku puzzles are given with answers. Students should start with puzzle 1 as this contains the most detailed explanation of how to solve it. Puzzle 3 is the most difficult.

Further puzzles can be readily generated using the Excel® templates in the post-16 sections *DIY Su Doku* CDROM index 35EX and *DIY ionic Su Doku* CDROM index 36EX. Instructions for doing this can be found on pages 71–74.

Polymer puzzles

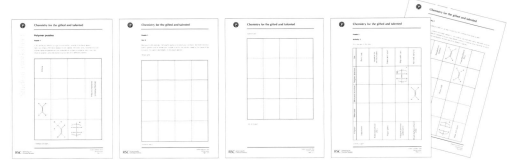

Student worksheet: CDROM index 16SW

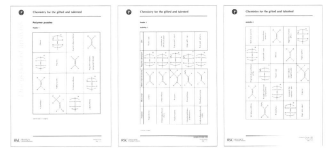

Discussion of answers: CDROM index 16DA

Topics

Names and displayed formulae of polymers and their associated monomers. Puzzle 2 also incorporates the uses of polymers.

Level

Able students in the 14–16 age range.

Prior knowledge

Names and displayed formulae of polymers and their associated monomers.

Rationale

This activity is designed to develop the students' higher order thinking – particularly critical thinking skills – in the context of problem solving. It should help students to recall the facts about some common polymers.

Use

These puzzles could be used as part of a topic on polymers or during revision. Puzzle 1 should be used first.

This page has been left blank intentionally.

Crude oil Su Doku

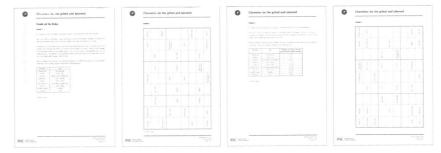

Student worksheet: CDROM index 17SW

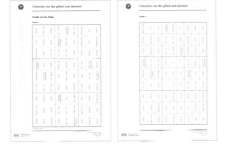

Discussion of answers: CDROM index 17DA

Topics

The fractions of crude oil, their uses and range of carbon chain length.

Level

Able students in the 14–16 age range.

Prior knowledge

Fractional distillation of crude oil.

Rationale

This activity helps the students interact with the information that they need to recall as they need to continually refer to that information to solve the puzzle.

Use

These puzzles could be used as part of a topic on oil or during revision. Puzzle 1 should be used first.

Further puzzles can be readily generated using the Excel® templates in the post-16 sections *DIY Su Doku* CDROM index 35EX and *DIY ionic Su Doku* CDROM index 36EX. Instructions for doing this can be found on pages 71–74.

Note

Another useful resource is the RSC online game *'Oilstrike'* ***www.rsc-oilstrike.org*** (accessed July 2007).

This page has been left blank intentionally.

Trends in reactivity in the Periodic Table

Student worksheet: CDROM index 18SW

Discussion of answers: CDROM index 18DA

Topics
Trends in reactivity of Groups 1 and 7 and the ionic bonding model.

Level
Very able students aged 14–16.

Prior knowledge
Ionic bonding and the Periodic Table.

Rationale
This activity aims to:
- help students develop a tool (flowcharts) to aid organisation of their line of reasoning;
- help students explore links between trends in the reactivity of Groups 1 and 7 and atomic structure;
- give students the opportunity to use and critically evaluate the relevance of ionisation energy data to the reactivity series of metals;
- reinforce the idea of chemical changes being driven by energy changes; and
- challenge the idea that Group 1 metals *want to give away* their outer shell electron.

Use
This could be used to follow up some work on the Periodic Table where the trends in reactivity in Groups 1 and 7 have been identified. It can be used as a differentiated activity for the more able students within a group.

When the students have completed the worksheet they should be given the *Discussion of answers* sheet. They could check their own work or conduct a peer review of the work of another student or group.

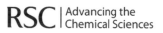

Volume changes

Volume changes

PowerPoint® presentation (Student worksheet): CDROM index 19SW

Volume changes
Answers to tasks 1 and 3
(the answers to tasks 2 and 4 will
vary from person to person)

PowerPoint® presentation (Discussion of answers): CDROM index 19DA

Topics
Metacognition, thinking tools, investigations and density.

Level
Able students aged 14–16.

Prior knowledge
Concentration of solutions, neutralisation reactions and ionic equations.

Rationale
This activity is designed to develop the students' metacognition (thinking about thinking). As they develop these skills they should have more thinking strategies with which to approach problems in chemistry and other subjects.

The activity uses a reflective method to develop metacognition. The students are asked to solve a problem and then reflect on the thinking tools or strategies that they used.

Use
This activity is best used with a whole class with teacher support. The PowerPoint® presentation (CDROM index 19SW) on volume changes should be shown to the class and should be moved on to the next slide only when the task has been completed and time allowed for discussion.

Students work in groups of three or four. They should only see one slide of the Powerpoint® presentation at a time, so, if this is used in paper form, the teacher could issue the sheets for tasks 1–4 as they are required. The aim is to get the students thinking about and discussing thinking styles, so it is important to allow plenty of time for discussion.

It is important to stress, on tasks 2 and 4, that we are interested in the thinking tools rather than the answers to the questions.

The students may need more examples of what is meant by thinking tools so the list *Thinking cards* (CDROM index 21TC) is included for teachers to give examples – please refer to *A question of thinking* on page 41. It is worth the teacher working through the activity and recording the thinking tools that they used as examples for the group.

A new kind of alchemy

Student worksheet: CDROM index 20SW

Discussion of answers: CDROM index 20DA

Topic
The Periodic Table.

Level
Able post-16 students.

Prior knowledge
Simple atomic structure and how the Periodic Table relates to electronic configuration.

Rationale
This presents some cutting edge research for post-16 students in a context that they can appreciate. It shows the students there are still big ideas to be explored in chemistry and should promote research as a career choice. The students are asked to speculate about questions where there are no known answers. This is designed to develop their creative thinking skills.

Use
This activity could be given at almost any stage in a post-16 chemistry course but is most appropriate in connection with topics about the Periodic Table. The activity could be done by a whole class but is also very well suited to use as a differentiated activity for the more able or more creative thinkers. When this was trialled, some less able students found it demoralising that there were not definitive answers to some of the questions.

This page has been left blank intentionally.

A question of thinking

Student worksheet: CDROM index 21SW

Discussion of answers: CDROM index 21DA

Thinking cards: CDROM index 21TC

Topics

Metacognition and thinking tools.

Level

Able post–16 students.

Prior knowledge

General chemistry background knowledge.

Rationale

This activity is designed to develop the students' **metacognition** (thinking about thinking). As they develop these skills they should have more thinking tools with which to approach problems in chemistry and other subjects.

Different thinking tools are described and given nicknames. The students have to decide which tools might be appropriate for different chemistry questions. This should encourage them to use a wider variety of thinking tools in their work.

Use

This can be used as a differentiated activity for the most able students in a mixed ability group, although there are benefits to using it with a whole class with teacher support.

Before the lesson the thinking tools need to be photocopied (preferably onto coloured card) and the sheets cut up into individual thinking tools. Each group of students needs a complete set of thinking tools (including blanks).

The students work in groups of three or four. They discuss how each thinking tool works and arrange them so that the creative ones are close together and the logical ones are close together. The students then consider a number of tough chemistry questions and choose appropriate thinking tools to tackle each question. They should not try to answer the question (in chemistry terms) at this stage.

Each group can challenge another to answer a question from the list using the thinking tools selected by the group. A competitive element could be added here with the team successfully answering the most questions from the list, demonstrating that they used the thinking tools they chose, winning a prize.

The students may be curious about the chemical answers to some of the questions and, if so, should be given the *Discussion of answers* sheets. The activity could be adapted for less able students by replacing some of the toughest questions with some easier ones.

The aim of this activity is to get the students thinking about and discussing thinking styles, so it is important to allow plenty of time for discussion. The activity takes around 1 hour 30 minutes to complete.

Covalent bonding

Student worksheet: CDROM index 22SW

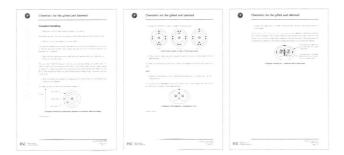

Discussion of answers: CDROM index 22DA

Topics
Energetic stability and molecular orbitals.

Level
Able post-16 chemistry students or extremely able students aged 14–16 who have covered covalent bonding in their pre-16 course.

Prior knowledge
Dot and cross diagrams and electron energy levels or shells (2, 8, 8 *etc*).

Rationale
The over-reliance on 'to get a full outer shell' as the concept underpinning bonding has been identified in previous RSC publications[1]. This activity seeks to develop an understanding of covalent bonding in terms of energetic stability rather than full shells.

Use
This activity can be used as an introduction to the further study of covalent bonding in post-16 chemistry courses, perhaps instead of revisiting dot and cross diagrams. It can be done as a differentiated activity with gifted 14–16 year olds to extend their understanding and challenge their thinking.

The students should be given the worksheets and asked to work through part 1. They should then be given the *Discussion of answers* sheet so that they can clarify the points raised in part 1. They should then move on to part 2.

Note

A word of caution: the energy levels drawn in the molecules do not represent the shape of the orbits or orbitals (in the same way that they do not for atoms). They are simply a way of representing the energy level of the electrons. There may be some students who misunderstand this point.

 This symbol means those questions are best tackled as a discussion if a group of students is doing this activity.

1   K. Taber, *Chemical misconceptions – prevention, diagnosis and cure*, London: Royal Society of Chemistry, 2002.

Curly arrows and stereoselectivity in organic reactions

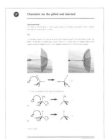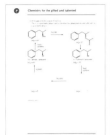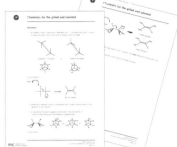

Student worksheet: CDROM index 23SW

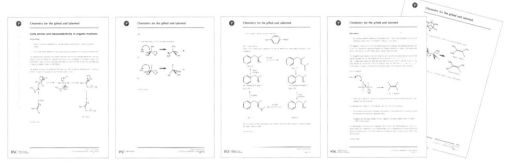

Discussion of answers: CDROM index 23DA

Topics

Curly arrows, the mechanism for esterification, inversion during S_N2 reactions, saw-horse representations and stereoselectivity in elimination reactions.

Level

Very able post-16 students.

Prior knowledge

Curly arrows in organic mechanisms, nucleophilic attack, electrophilic attack and optical isomerism.

Rationale

'All practising chemists protect themselves from being crushed by the vastness of organic chemistry by moulding it and ordering it with curly arrows. Without curly arrows, chemistry is chaos, and impossible to learn. Curly arrows unify chemistry, and are essential to the solution of problems'[1].

Organic chemistry is sometimes presented to students with the emphasis on facts to recall rather than the underlying principles. Very able students tend to be interested in fundamental concepts and put off by a 'just learn it' approach. This activity explores some of the intellectually satisfying aspects of organic chemistry. The first part develops the use of curly arrows. If you have some very able students in a group, then there is a ready made opportunity for differentiation by giving them a chance to draw curly arrow mechanisms for whatever reactions they meet.

Evidence from the UK qualifying competition for the *International Chemistry Olympiad* suggests that even very able students get little practice in presenting the three-dimensional shapes of molecules. This activity gives the students an opportunity to develop those skills.

Use

This activity requires a good background knowledge of organic chemistry and should be used with students towards the end of their organic chemistry course. It can be given to the most able students in a group to work through independently. They should also be given the *Discussion of answers* sheet to review their progress. Students should be able to assemble simple 3D molecular models (such as Molymod®) when they are doing this activity.

[1]   Clayden, Greeves, Warren and Wothers, *Organic Chemistry*, Oxford: Oxford University Press, 2001.

Entropy and equilibrium

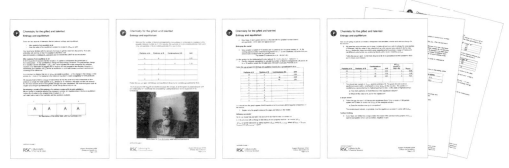

Student worksheet: CDROM index 24SW

Discussion of answers: CDROM index 24DA

Alternative question 11: CDROM index 24AQ

Alternative answer 11: CDROM index 24AA

Excel® spreadsheets: CDROM index 24E1, 24E2, 24E3, 24E4, 24E5 and 24E6

Topics

Entropy and equilibrium.

Level

Gifted post-16 students who are motivated about understanding the fundamental concepts of entropy and equilibrium.

Prior knowledge

Entropy, the second law of thermodynamics (expressed in terms of ΔS_{total}), calculations of permutations, factorial function (!), natural logarithms, entering formulae in spreadsheets and dragging and filling-in spreadsheets.

Other concepts

$S = k\ln W$ and modelling.

Rationale

Chemistry at its best is an interconnected web of concepts, skills and facts. Very able students find the linking of concepts particularly rewarding. This activity shows the students the fundamental link between entropy and equilibrium and increases their understanding of scientific models. It highlights the importance of mathematical descriptions in physical chemistry. They also use and develop their maths and IT skills.

The activity focuses on entropy, rather than Gibbs energy, because those who have gone as far as Gibbs energy should still be able to follow a discussion based on entropy. Entropy is a more fundamental concept than Gibbs energy.

Use

This worksheet is designed to be used by students working individually at a computer, or during a teacher-led group discussion via a projector. Teachers may find the spreadsheets helpful (calculation 1 in particular) for use in their general teaching.

There are references in the worksheet to direct students to various Excel® spreadsheets stored on the CDROM. When they have finished working on the spreadsheets students should close the windows containing the spreadsheets – they should **not save changes** to the spreadsheets! The 100 particle model spreadsheets are set up with protection on, but more accessible sheets are available if students want to see the workings.

A more detailed alternative to question 11 is the *UK 2004 Round 1 Olympiad selection paper question 5* which is included (CDROM index 24AQ and 24AA).

Further reading

A more complete explanation of the connection between the entropy of mixing and equilibrium can be found in:

> Keeler & Wothers, *Why chemical reactions happen*, Oxford: Oxford University Press, 2003 (distributed to secondary schools and colleges by the RSC Spring 2006). Chapter 8 should be used first.

Formal charge

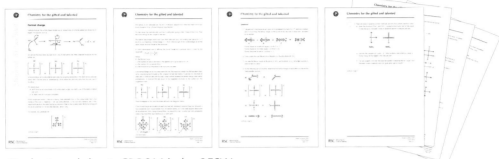

Student worksheet: CDROM index 25SW

Discussion of answers: CDROM index 25DA

Topics

Transition state, resonance structures, reactive intermediates, carbocations and electronegativity.

Level

Very able post–16 students.

Prior knowledge

Dot and cross diagrams, dative bonds, oxidation numbers and curly arrows in organic mechanisms.

Rationale

This activity introduces formal charge – a useful tool which otherwise might not be taught. The formal charge model treats bonds as pure covalent, in contrast to the oxidation state model which treats bonds as ionic. These two models should be seen as opposite ends of a continuum with the real charge on atoms being somewhere in between the two.

This activity explores the usefulness of the formal charge model and gives the students the experience of refining a model when it starts to fail and realising situations when it should be abandoned in favour of other models. It encourages the students to think of quantities like formal charge and oxidation number as useful models with artificial rules that do not represent accurately the real electron density distribution in molecules. The students are asked to suggest rules to help cope with transition metals – this should help them appreciate the design of the rules they use for assigning oxidation numbers.

It gives a good rationale for the charges in organic mechanisms. The students should uncover the link between formal charge and dative bonds.

Use

It can be used as a differentiated activity for the most able in a mixed ability group or for a whole class with teacher support. If used with a whole class, rather than getting them to wade through the lengthy instructions, it would be good to go through a few examples together so that they quickly get a feel for how it works. Less able students found the instructions a lot to read and digest without support.

It could be done in two parts, part 1 being the introduction and questions 1–5 with part 2 being the remainder.

 This symbol means those questions are best tackled as a discussion if a group of students is doing this activity. It is intended that written answers are not required for these questions.

Question 9 is included mainly for those who want some more background.

Mixing drinks

PowerPoint® presentation (Student worksheet): CDROM index 26SW – see Use *below.*

Discussion of answers: CDROM index 26DA

Topics
Metacognition, thinking styles, investigations, trends and entropy changes in solution.

Level
Able post-16 students.

Prior knowledge
Intermolecular forces, hydrogen bonding and the structure of ice.

Rationale
This activity is designed to develop the students' metacognition (thinking about thinking). As they develop these skills they should have more thinking strategies with which to approach problems in chemistry and other subjects.

The activity uses two methods to develop metacognition. First, the students are asked to solve a problem and then reflect on the thinking styles that they used. In the other method students attempt to answer some questions and then discuss four modelled thinking styles of fictional students.

Use
This can be used as a differentiated activity for the most able in a mixed ability group, although there are benefits to using it as a whole class activity with teacher support.

You need a photocopy of the four students' (A-D) thinking styles for each group. It is best to show the early slides, up to and including the instructions for task 4, using a projector.

The students work in groups of four. The students should only see one slide at a time, so if this is used in paper form, the teacher can give out the sheets for tasks 1–3 as they are required. For task 4 the modelled thinking styles of students A–D should be printed out (on coloured paper if you want to reuse them) and each group of students given one set of the modelled thinking styles (one copy of each student A-D). The students in the group then each take one thinking style for task 4.

Student D uses entropy in their thinking, but their thinking **style** can still be described even if the topic of entropy has not been covered.

The aim is to get the students thinking about and discussing thinking styles, so it is important to allow plenty of time for discussion.

Which is the odd one out? (Organic)

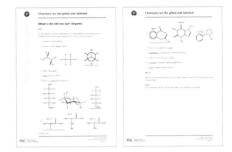

Student worksheet: CDROM index 27SW

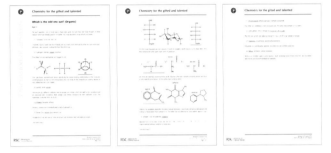

Discussion of answers: CDROM index 27DA

Topics

Oxidation level, functional groups, saw-horse projections, drug molecules, amino acids, addition and condensation polymers and the ring and chain forms of glucose.

Level

Able post-16 chemistry students.

Prior knowledge

Chirality, oxidation level, electrophiles, nucleophiles, bases and π bonds.

Rationale

This activity should encourage a rapid consideration of the range of concepts met in organic chemistry. As scientists we survey the available models or concepts to decide which is most pertinent to the current problem. This activity is designed to develop those skills. The students need to think laterally in some cases. It may help students develop the skills needed for synoptic style questions.

Use

The activity can be used as a revision tool towards the end of a post-16 course. It is synoptic in nature and draws on several aspects of organic chemistry. It is probably best used as a discussion tool in small groups.

This page has been left blank intentionally.

Olympiad past paper questions

Student worksheet: CDROM index 28SW

Discussion of answers: CDROM index 28DA

Topics

Topics met in post-16 chemistry

Level

Post-16 students.

Prior knowledge

The content of post-16 courses.

Rationale

The questions on the CDROM (index 28SW) are a selection of the questions used in Round 1 of the selection process for the *International Chemistry Olympiad*. They are accompanied by a *Discussion of answers* (index 28DA).

Further examples are to be found at ***www.rsc.org/olympiad*** (accessed July 2007).

Use

For revision or for preparation for the *Olympiad* selection procedure.

This page has been left blank intentionally.

Organic reaction maps

Organic reaction maps: CDROM index 29RM

PowerPoint® drag and drop version: CDROM index 29DD

Topics
Oxidation level, types of reaction and functional groups. alkenes, alkanes, alcohols, aldehydes, ketones, halogenoalkanes, amines, carboxylic acids, esters, acyl chlorides, acid anhydrides, nitriles and amides.

Level
Post-16 students of all abilities.

Prior knowledge
A complete course in organic chemistry, apart from aromatic compounds.

Rationale
This activity encourages the use of mind maps to organise information. It also highlights where oxidation and reduction are involved in transformations between functional groups. The students should end up with an overview of the reactions of aliphatic compounds. This should help the students plan simple synthetic pathways. The activity highlights the logic of organic chemistry by showing that in the great variety of reagents and conditions there are still only a few different types of reaction going on.

Use
This could be done towards the end of a post-16 course, to summarise the non-aromatic organic chemistry covered, or as revision before the examinations.

1. **Organic reaction maps** *(CDROM index 29RM)*
   In this version the students work on paper. The outline concept map is best photocopied onto A3 paper to give the students room to annotate fully.

The activity could be done as a group, with a master on an OHP or projected, and suggestions for links coming from the class. It can also be done individually and is produced with different levels of scaffolding support.

This activity is suitable for all abilities in different forms. Weaker students can do tasks 1, 2 and 3 (or the drag and drop version on the computer). Stronger students can go straight to task 4.

The most able students could go on to devise a mind map showing the interrelation of different functional groups with a rationale behind where the compounds are put on the page.

The compounds might be organised by:
- the extent of oxidation (working from the centre out); or
- whether they are susceptible to nucleophilic or electrophilic attack *etc*.

Teachers may want to adapt the tasks and answer sheet to fit their awarding body specification more closely. The same tasks are available with the structures given rather than the names. The answers are on slides within the slide show.

Although written in PowerPoint®, the organic mind map is not intended for use as an interactive presentation; it is better to use the drag and drop version if you want an interactive presentation. Teachers may want to show individual slides to discuss what the students are doing on paper.

2. PowerPoint® drag and drop version (*CDROM index 29DD*)
The drag and drop version is for students to use working at a computer.

For the drag and drop feature to work the students need to select *view show*. With drag and drop operating, several of the normal navigation features in PowerPoint® are suspended. Students navigate through the reaction map tasks by left clicking the appropriate links in the corners of the screen, including to check their own answers.

The 'home' button at the top left of every slide links to an index page.

To close the slide show without saving the changes, the students need only to left click *close presentation* at the top right of any slide.

To drag a coloured box you need to select it by *single left mouse clicking and releasing*, moving the mouse until the object is where you want it to be and then *left click and release* again to drop the object.

If you experience problems with dropping, and there is a black border either to the sides or top and bottom of the screen, then you need to change the screen resolution so the slide fills the screen. With all windows closed, right click on the background then select *properties – settings* and choose a screen resolution that fills the screen.

There are three tasks in the drag and drop version:
- Labelling the reaction types
- Matching structures to names
- Adding reagents and conditions to reactions

In each case the students drag the appropriate coloured box to the correct position. The answers are on slides within the slide show.

Note about macros in drag and drop version *(CDROM index 29DD)*

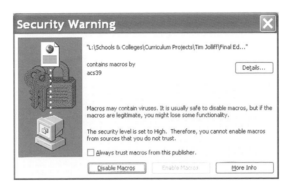

It this security warning dialogue box appears when accessing the drag and drop version, you should click *Enable Macros* if this option is available. If it is unavailable you should first click the *Always trust macros from this publisher* box and then click on *Enable Macros*.

This page has been left blank intentionally.

Oxidation numbers

Student worksheet: CDROM index 30SW

Discussion of answers: CDROM index 30DA

Topics
Working out oxidation numbers from electronegativity values, challenging redox questions and comparing the two methods of assigning oxidation numbers – electronegativity values and oxidation number rules.

Level
Very able post-16 students.

Prior knowledge
Polar bonds and electronegativity.

Rationale
Part 1 introduces oxidation numbers by giving a conceptual foundation for them in terms of electron accounting and polar bonds. Very able students tend to prefer this approach to starting with a set of rules.

Part 2 gives some examples of the usefulness of oxidation numbers in working out balanced equations.

Part 3 shows where the bond polarity approach gives greater insight into the oxidation numbers of atoms in particular examples. It then shows how the model used so far needs refining and asks the students to think about how this might be done.

This should lead to a greater understanding of the nature of the models that we use in chemistry as well as methods of assigning oxidation numbers. The activity shows where non-integer oxidation numbers arise and what they mean.

Use

The activity should be done in two sections:

- Part 1 is the introduction to oxidation numbers. It is written so that students can be given this activity as a first introduction but it can also be used when they have met the oxidation number rules. It is suitable for students of all abilities.
- Parts 2 and 3 are the follow up to class work on assigning oxidation numbers using the usual rules and constructing redox equations. These parts are suitable for very able students.

Students should be given the *Discussion of answers* section to the part they have completed before moving on to the next part.

 This symbol means those questions are best tackled as a discussion if a group of students is doing this activity. It is intended that written answers are not required for these questions.

The note on Al_2Cl_6 seeds the thought that not all bonds are two-electron two-centre bonds and some students may want to research that more. A brief description can be found at *http://en.wikipedia.org/wiki/Three_center_bond* (accessed May 2007).

Rates and equilibria

Student worksheet: CDROM index 31SW

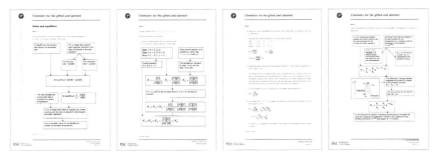

Discussion of answers: CDROM index 31DA

Topics

Equilibria (K_c, K_p, Le Chatelier's principle), the rate equation, the rate constant, the Arrhenius equation and activation energy.

Level

Able post-16 students.

Prior knowledge

Rates of reaction, rate equations and K_c (K_p would be helpful but it is explained in the worksheet).

Rationale

This activity demonstrates the links between the topics of rates of reaction and the equilibrium law. It provides students with an explanation of the equilibrium law and helps them explain why Le Chatelier's principle works for temperature, concentration and pressure.

The information card approach helps students develop their communication skills and their synthesis – higher order thinking.

Use

This is a group activity and groups of three or four students are ideal. Each group should have one set of the information cards which is dealt out between the members of the group. The cards are labelled with which part of the activity they are for; they can all be dealt out at the start and the students told to refer just to the cards for the part of the activity they are working on. They should not 'lay their cards out on the table' unless they are stuck.

Cards which have an asterisk (*) can be held back by the teacher initially and given to the students when or if they need help.

The groups could produce one set of written answers which could be reviewed in the light of the *Discussion of answers* sheet by another group (with instructions to be positive in their comments).

If you want to reuse the information cards it is helpful to reproduce them on different coloured card or paper.

RSC | Advancing the Chemical Sciences

The second law of thermodynamics

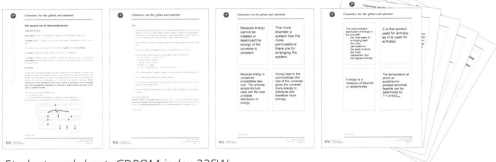

Student worksheet: CDROM index 32SW

Discussion of answers: CDROM index 32DA

Topics

System, surroundings, disorder, spontaneous changes, entropy and the second law of thermodynamics.

Level

More able post-16 students.

Prior knowledge

Enthalpy changes and ΔH.

Rationale

Entropy is a topic that very able post–16 chemistry students should meet irrespective of whether it forms part of their course specification or not. Gifted students are often attracted by the fundamental ideas and the second law of thermodynamics is both fundamental and surprising the first time you meet it. This activity aims to introduce the topic in a way that uses the students' synthesis skills to piece together several pieces of information. Answers are available and the students explain their understanding to one another thus reinforcing their ideas.

Use

This is a group activity best used with groups of three to five students, but could be done by two or even a single student – with some adaptation. If entropy is part of the course then this activity can be used as an introduction to the topic with the whole group or a summary exercise at the end. If used as an introduction with a mixed ability group, trialling suggests that they may not assimilate all the information, but that it will stimulate questions and interest.

A website that might be used to introduce some of the terms is *www.mhhe.com/physsci/chemistry/chang7/esp/folder_structure/en/m4/s1/index.htm* (accessed May 2007). If used as extension work for the most able students, it may not need much input from a teacher. It may also stimulate sufficient interest for them to read up on the topic.

The students should be split up into groups of three to five. They should each be given the worksheet and one set of information cards per group. If you plan to reuse the cards they are best produced in a different colour for each set. Once they have arranged the cards into a reasoned order they should be given the example arrangement of information to aid their explanation to another student. They can then tackle the remainder of the questions.

The *Discussion of answers* sheet should be given so they can review their own or each others' ideas.

 This symbol means those questions are best tackled as a discussion if a group of students is doing this activity. Written answers are not expected. Some of the questions asked towards the end of the activity are quite involved and may be best tackled in a separate lesson once the main ideas have sunk in.

Shapes of molecules and ions

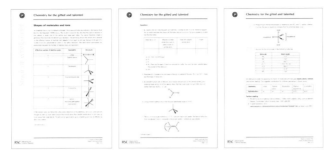

Student worksheet: CDROM index 33SW

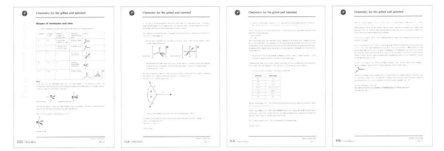

Discussion of answers: CDROM index 33DA

Topics

Valence shell electron pair repulsion (VSEPR) theory application.

Level

Very able post-16 students.

Prior knowledge

Valence shell electron pair repulsion (VSEPR) theory.
(An explanation of the VSEPR theory is given but this activity is best done after some discussion of the theory in class.)
Resolving forces (questions 3 and 4).

Rationale

This activity draws some extra concepts and mathematical skills into the discussion of molecular shape. A distinction between geometry around the central atom and the shape of the molecule is made. Students are asked to apply the VSEPR theory to some challenging examples, including species with unpaired electrons. Students are introduced to the alternative model of hybridised atomic orbitals.

Use

This can be used as a follow up to a theory lesson on VSEPR theory for very able students. It can also be used as a differentiated activity for the most able students in a mixed ability group. If the students do not know how to resolve forces, they should leave out questions 3 and 4.

This symbol means those questions are best tackled as a discussion if a group of students is doing this activity.

Swimming pool chemistry

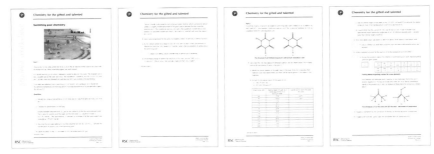

Student worksheet: CDROM index 34SW

Discussion of answers: CDROM index 34DA

Topics
Part 1 Equilibria (K_c, K_{sp}, K_a, buffers, Le Chatelier's principle) and entropy (solvent effects).

Part 2 Mass spectra (quite involved spectra to interpret), oxidation numbers, curly arrow mechanisms, nuclear magnetic resonance (NMR) spectroscopy, delocalisation, tautomerism and relative molecular mass calculations.

Level
Post-16 students who are close to the end of their chemistry course.

Prior knowledge
K_c, pH, Le Chatelier's principle, buffers, spectra, NMR and curly arrows.

Rationale
This activity is synoptic in nature, with questions on many different aspects of chemistry. It is context driven *ie* looks at a collection of interesting chemistry that can be derived from swimming pools. Some extension work is introduced – *eg* keto-enol tautomerism.

Use
The activity comes in two parts and these can be given separately to the students. Part 1 can be used at the end of teaching equilibrium or at the end of a course. Part 2 should be attempted at the end of a course.

 This symbol means those questions are best tackled as a discussion if a group of students is doing this activity. A written response is not normally required.

This page has been left blank intentionally.

DIY Su Doku

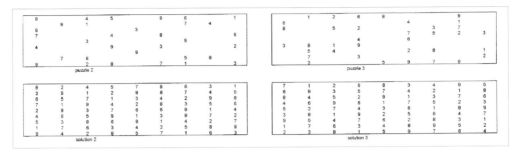

Su Doku DIY template: CDROM index 35EX

Whenever there is a list of related facts that you want your students to learn, a Su Doku puzzle may be an appropriate tool to help. For example, it could be used for learning the formulae of a list of named compounds, the uses of the fractions of crude oil or the test for ions. You may want to generate new chemistry Su Doku puzzles and this spreadsheet makes that quick and easy.

To generate new Su Doku puzzles you need to open the Excel® file called *Su Doku DIY template* (index 35EX).

A) To create puzzles that contain alternative entries within the grid:

Go to the third tab (*alternative reference data*) and copy one of the tables of data (9x9 cells) – eg for the table with columns *material, property* and *use,* copy from 'glass' (top left hand cell) to 'tyres' (bottom right hand cell).

Go to the first tab (*3 columns of data puzzle*) and paste the copied table of data onto the 9x9 cell table entitled *Reference data* (top right hand table). The new puzzle will appear in the table to the left of the *Reference data* table.

To generate the solutions, go to the second tab (*3 columns of data solution*) and paste the same copied table onto the 9x9 cell *Reference data* table (top right hand table). The solutions will appear in the table to the right of the *Reference data* table.

B) To rearrange the same entries within a grid:

Go to the fourth tab (*alternative number Su Doku*) and copy one of the 9x9 cell 'puzzle' tables (at the top of the spreadsheet).

Go to the first tab (*3 columns of data puzzle*) and paste the copied table onto the top 9x9 cell table of numbers. The new puzzle will appear in the table above the number table.

To generate the solutions, return to the fourth tab (*alternative number Su Doku*) and copy the corresponding 9x9 cell 'solution' table (below the chosen 'puzzle' table).

Go to the second tab (*3 columns of data solution*) and paste the copied 'solutions' table onto the top 9x9 cell table of numbers. The solutions will appear in the table above the numbers table.

Note

The *Reference data* table needs three columns of data. If you wish to use two columns of related data (rather than three):

Produce a 2x9 cell table, with the two related variables in columns 1 and 2 respectively.

Paste the information from column 1 into column 3 to complete the 3x9 cells (*ie* columns 1 and 3 will contain the same data). For an example, see the top left hand table in the third tab (*alternative reference data*).

Proceed as before.

If you wish to write your own Su Doku entries, follow the format on the third tab (*alternative reference data*) to create your data table, then follow the procedures above.

Once you have produced your puzzle and solutions table, copy and paste them into new Excel® spreadsheets, ready to print for your students.

DIY ionic Su Doku

NaI	Na2T			FeCl2			Cs2O	CsR
NaR	Na2O		FeT		FeX		CsSZ	CaF
		FeR2				Cs2T		
		NiT	SnCl4		SnR4	CaF2		
	NiF2						CaX	
		NiCl2	Sn3Z4		SnO2	CeI2		
						VR5		
Fe2X3	FeR3		AgQ		AgCl		VF5	VI6
FeF3	FeCl3			AgR			V2T5	VQ5

	Na+	Fe2+	Cs+	NiQ+	Sn4+	Ca2+	Fe3+	Ag+
0								
1	NaCl	FeCl2	CsCl	NiCl2	SnCl4	CeCl2	FeCl3	AgCl
2	NaF	FeF2	CsF	NiF2	SnF4	CaF2	FeF3	AgF
3	NaI	FeI2	CsI	NiI2	SnI4	CaI2	FeI3	AgI
4	Na2O	FeO	Cs2O	NiO	SnO2	CeO	Fe2O3	Ag2O
5	Na2X	FeX	Cs2X	NiX	SnX2	CeX	Fe2X3	Ag2X
6	Na2Z	FeZ2	Cs2Z	Ni2Z	Sn2Z	Ca2Z	FeZ	Ag2Z
7	NaQ	FeQ2	CsQ	NiQ2	SnQ	CaQ2	FeQ3	AgQ
8	NaR	FeR2	CsR	NiR2	SnR4	CaR2	FeR3	AgR
9	Na2T	FeT	Cs2T	NiT	SnT2	CaT	Fe2T3	Ag2T

Reference data

3	9	0	0	1	0	0	4	8
8	4	0	9	0	5	0	6	2
0	0	0	8	0	0	9	0	0
0	0	9	1	0	8	2	0	0
0	2	0	0	0	0	0	5	0
0	0	1	6	0	4	3	0	0
0	0	0	0	0	0	8	0	0
5	8	0	7	0	1	0	2	3
2	1	0	0	8	0	0	9	7

Ionic Su Doku DIY template: CDROM index 36EX

You may want to generate different or further ionic Su Doku puzzles and this spreadsheet makes that quick and easy.

To generate new Su Doku puzzles you need to open the Excel® file called *Ionic Su Doku DIY template* (index 36EX).

A) To create puzzles that contain alternative entries of your choice within the grid:

Go to the first tab (*template*) and replace the data in the cells of the *Reference data* table (9x9 cells) with your chosen alternative data. The new puzzle will to appear to the left of the *Reference data* table.

To generate the solutions, copy the new *Reference data* table (9x9 cells) in the first tab (*template*).

Paste the copied table onto the 9x9 *Reference data* table in the second tab (*answer sheet*). The solutions will appear to the left of the *Reference data* table.

Note
You may find that formatting the formulae (for the *Reference data* table) is easiest in Word® which can then be pasted into Excel®. Also, subscripts may not work in Excel®.

B) To rearrange the same entries within a grid:

Go to the third tab (*alternative number Su Doku*) and copy one of the 9x9 cell 'puzzle' tables (at the top of the spreadsheet).

Go to the first tab (*template*) and paste the copied table onto the 9x9 cell table of numbers. The new puzzle will appear in the table above the number table.

To generate the solutions, return to the third tab (*alternative number Su Doku*) and copy the corresponding 9x9 cell 'solution' table (below the chosen 'puzzle' table).

Go to the second tab (*answer sheet*) and paste the copied 'solutions' table onto the 9x9 cell table of numbers. The solutions will appear in the table above the numbers table. Once you have produced your puzzle and solutions table, copy and paste them into new Excel spreadsheets, ready to print for your students.